ZOOM

Dinosaurs

BLACKBIRCH®
PRESS

THOMSON
★
GALE

San Diego • Detroit • New York • San Francisco • Cleveland
New Haven, Conn. • Waterville, Maine • London • Munich

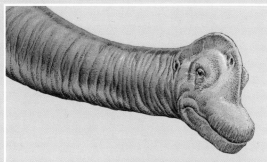

CONTENTS

LET'S ZOOM!

When you use the zoom feature on a camera, you bring images from the distance into close-up focus without moving yourself. For example, you can capture the image of a butterfly on a leaf while keeping your distance. This book works in exactly the same way.

Imagine you were able to travel back in time 100 million years and point a camera at the whole Earth. The illustration on pages 6 and 7 shows what you would see in your viewfinder. (Earth looked very different then). Now zoom in to one part of the scene. You may find yourself looking at a landscape of erupting volcanoes and steaming, swampy forests. There are no signs of cities, roads, or farmland of course—only wild nature. Keep zooming, and you will eventually arrive in the unforgettable world of the most awesome creatures that ever walked on this planet: the dinosaurs. As you zoom in further, you'll witness a titanic contest between a giant, long-necked dinosaur and its terrifying predators. You'll make out much smaller inhabitants of the dinosaur world as they dart through the undergrowth. You'll discover a nest hidden away in that undergrowth, an egg, and a tiny dinosaur embryo inside that egg.

It's a fascinating journey, yet you will not have to move one inch! And you'll discover some amazing things about dinosaurs that only this incredible *zooming* book can show you . . .

JURASSIC EARTH

About 150 million years ago, during the Jurassic period, Earth's continents were much closer together than they are today. Only narrow, shallow seas separated Europe, North America, and Asia. South America, Africa, Antarctica, and Australia were all joined together in a "super-continent" known as Gondwana.

How do continents move? Earth's surface is divided into a number of slabs, called plates. Driven by currents in the molten rock deep inside Earth, they gradually slide around the globe. Whole continents may drift thousands of miles, colliding with others or splitting apart.

ZOOM IN TO A JURASSIC LANDSCAPE ON EARTH

LANDSCAPE

Just as the shapes of Earth's continents and oceans were very different millions of years ago, so too were Earth's landscapes. Worldwide, the climate was warm and humid. Moist winds from the oceans brought rain to inland regions.

EARTH'S CONTINENTS WERE MUCH CLOSER IN JURASSIC TIMES

Higher temperatures melted the polar ice, raising sea levels above where they are today. Active deep-sea volcanic eruptions along the cracks in Earth's crust raised the sea floor, making sea levels even higher.

Swampy lowlands covered great tracts of land. A typical Jurassic landscape was lush and green.

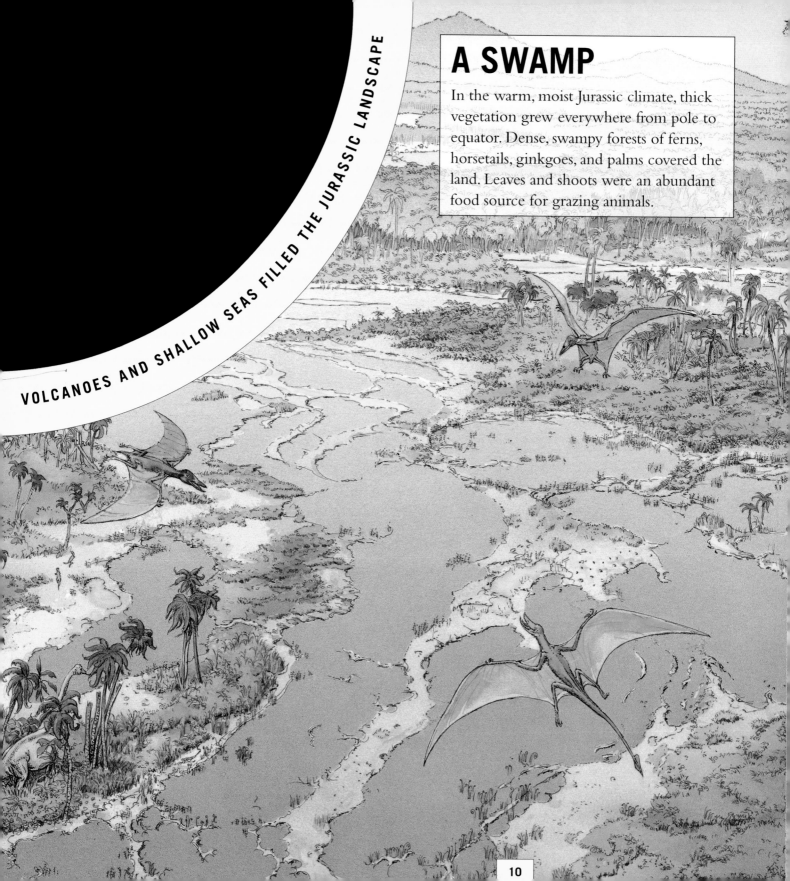

A SWAMP

In the warm, moist Jurassic climate, thick vegetation grew everywhere from pole to equator. Dense, swampy forests of ferns, horsetails, ginkgoes, and palms covered the land. Leaves and shoots were an abundant food source for grazing animals.

Pterosaurs—flying reptiles—circled above the land. They were especially found near coasts, where fish, their main source of food, was plentiful. Pterosaurs' wings were made of sheets of skin that linked their fourth fingers to their bodies. Their toothed beaks were perfect for plucking fish out of the water and carrying them away.

ZOOM DOWN TO GROUND LEVEL IN THE SWAMP

DINOSAURS

Some of the largest animals ever to walk on Earth lived during the Jurassic period. The sauropods *(see page 26)* were a group of plant-eating dinosaurs that grew up to 100 feet (30 m) in length. Some could bite off leaves from trees 40 feet (12 m) high. Rearing up on their hind limbs, they could reach higher still.

Cetiosaurus

Brachiosaurus

DENSE JURASSIC SWAMPS SWARMED WITH LIFE

Cetiosaurus

Eustreptospondylus

Cetiosaurus was a very sturdy sauropod. It was nearly 60 feet (18 m) long with pillar-like legs. Its spoon-shaped teeth were perfect for tearing leaves from trees.

Sauropods lived alongside many kinds of dinosaurs during the Jurassic. Smaller plant eaters browsed on ferns and other low plants. Flesh eaters, called theropods, were also abundant. Eustreptospondylus was a 23-foot (7 m) long killer. It specialized in swift gang attacks on lone sauropods.

ATTACK!

Although they were much smaller dinosaurs, several eustreptospondylus might launch an attack together against a massive cetiosaurus. These predators would circle their victim before attacking.

14

Whatever cetiosaurus lacked in speed, it made up for in sheer size. It could rear up on its back legs and crash down on its attackers. It might also flick its whip-like tail with great force into a predator's face, which could be a fatal blow. If they could avoid a cetiosaurus counterattack, the eustreptospondylus gang would rush at their prey, using their claws and dagger-like teeth to bring it down.

ZOOM IN CLOSER TO THE GROUND BENEATH THE DINOSAURS' FEET

SMALLER DINOSAURS

Not all dinosaurs were lumbering giants or powerful killers. Some, like compsognathus, were swift, cat-sized creatures that hunted for lizards, insects, or small mammals in the undergrowth. This tiny, slim dinosaur had thin legs, bird-like feet, and looked similar to archaeopteryx, one of the earliest-known birds. The two animals actually lived in the same part of the world at the same time, and may have been closely related.

GANGS OF SMALL DINOSAURS ATTACKED LARGE DINOSAURS

Compsognathus was quick on its feet. It may have been able to outrun lizards or catch dragonflies in flight.

Archaeopteryx

Compsognathus

Archaeopteryx was about the size of a chicken. Like its probable ancestors—flesh-eating dinosaurs—it had teeth and a long, bony tail. It also had feathers and wings. It could probably fly—or at least glide from tree to tree. Archaeopteryx may have fed on insects.

SMALLER DINOSAURS INHABITED THE JURASSIC UNDERGROWTH

DINOSAUR NEST

Here, some baby dinosaurs have hatched, fully formed, and already on their feet in search of food. Like birds and most reptiles, many dinosaurs made nests for their eggs. Some dinosaur mothers stayed by the nest to protect the eggs and the tiny hatchlings. Despite best efforts, however, the eggs (and the hatchlings) were an easy source of food for small mammals and other dinosaurs.

ZOOM CLOSER AND GO INSIDE A DINOSAUR EGG

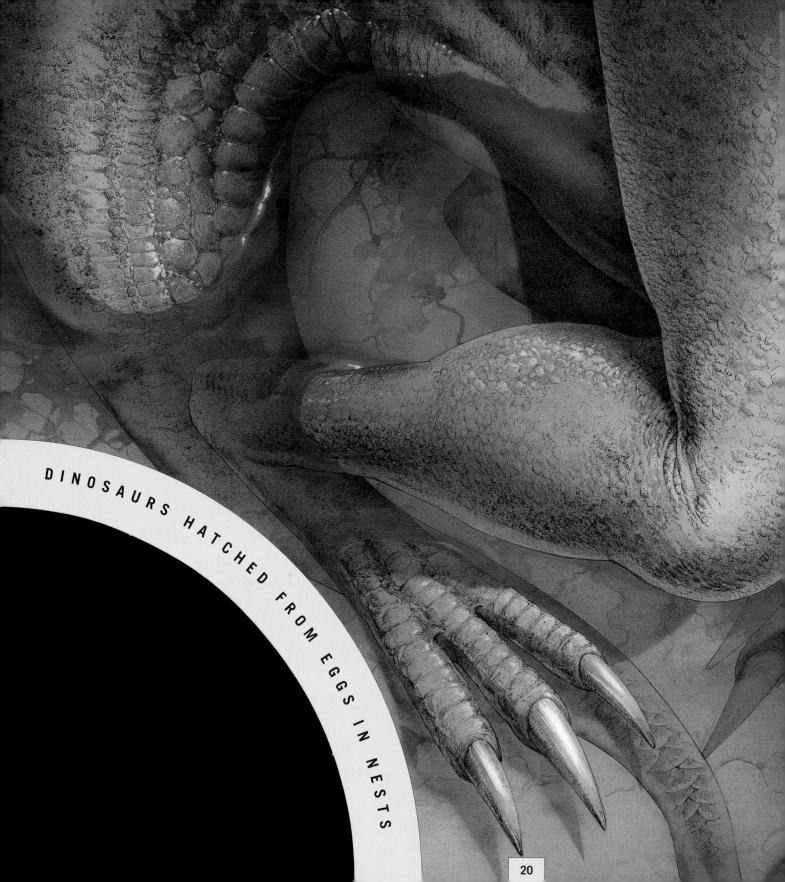

DINOSAURS HATCHED FROM EGGS IN NESTS

INSIDE AN EGG

Dinosaur eggs were hard-shelled, but they had tiny holes in the walls to allow the baby inside to breathe. When it was ready to emerge, a hatchling simply knocked the top off and climbed out. The hatchlings of certain dinosaurs were not fully developed. These newborns depended on their parents for care. Other species (like the megalosaurus illustrated) were born fully formed and able to fend for themselves immediately.

ZOOM DOWN TO THE SURFACE OF THE BABY DINOSAUR'S SKIN

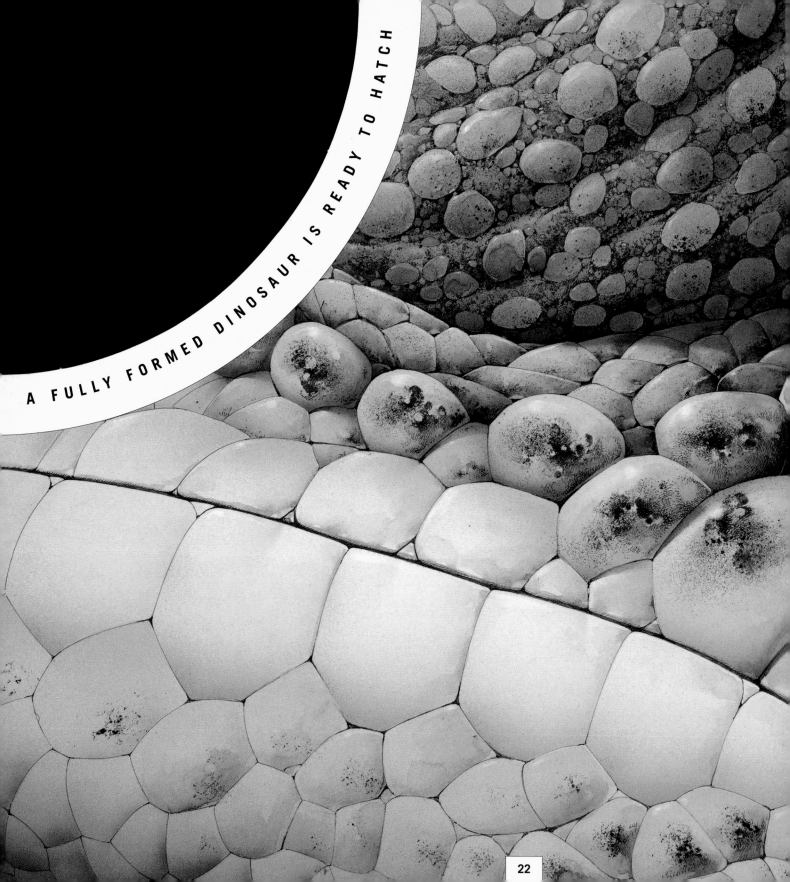

A FULLY FORMED DINOSAUR IS READY TO HATCH

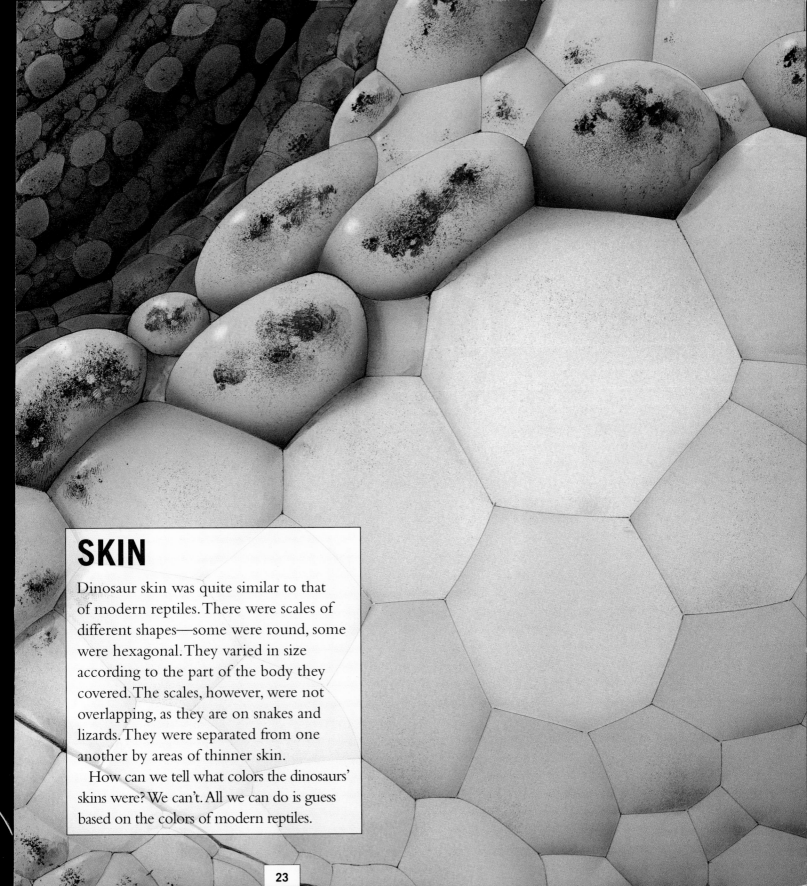

SKIN

Dinosaur skin was quite similar to that
of modern reptiles. There were scales of
different shapes—some were round, some
were hexagonal. They varied in size
according to the part of the body they
covered. The scales, however, were not
overlapping, as they are on snakes and
lizards. They were separated from one
another by areas of thinner skin.

How can we tell what colors the dinosaurs'
skins were? We can't. All we can do is guess
based on the colors of modern reptiles.

The ancestors of the dinosaurs basked in the swamplands of the Carboniferous forests. Heavy, lumbering amphibians—some more than 7 feet (2 m) long, still spent much of their time in the water. Living alongside the giant dragonflies and centipedes was the tiny reptile hylonomus.

Dendrerpeton (amphibian)

Hylonomus

DINOSAUR ORIGINS

Dinosaurs were reptiles that lived 230 to 65 million years ago, during the Triassic, Jurassic, and Cretaceous periods. Unlike other reptiles, they walked upright on legs held beneath their bodies, like birds and mammals. For more than 160 million years, dinosaurs of many kinds and sizes dominated life on land until they, along with marine and flying reptiles, became extinct (see page 28).

The first reptiles had evolved about 320 million years ago, during the Carboniferous period. They were descended from **amphibians**, animals with fish-like heads and tails but with four legs instead of fins. In the hot, swampy forests that blanketed the lowlands of Europe and North America at this time, amphibians multiplied quickly.

Dimetrodon lived during the Permian period. It was a pelycosaur—not a dinosaur. Its sail may have helped control its body temperature.

Although capable of living on land, amphibians stayed close to water, where (as frogs and newts do today) they laid their jelly-covered eggs and the young swam like fish. Eventually, some creatures found a way to lay hard-shelled eggs on land. These animals became the first **reptiles**.

The tropical Carboniferous forests gave way to the dry scrublands of the Permian and Triassic periods. This drying out favored the reptiles with their ability to lay eggs on land. Reptiles also evolved stronger jaw muscles, which enabled them to eat tough desert plants.

Two of the main reptile types included the mammal-like reptiles (from which mammals would later evolve) and the **archosaurs**. Equipped with powerful jaws and bony armor, the archosaurs became dominant during the Triassic period. Early archosaurs had a sprawling gait, but later in the Triassic, some kinds began to stand more upright. About 230 million years ago, the first dinosaurs, also from the archosaur group, had evolved. They were able to run around on their hind legs, which left their arms free to grasp prey. Among the early

Modern lizard

Early archosaur

Dinosaur

Dinosaurs were land reptiles. Neither marine reptiles nor flying reptiles were dinosaurs, although they lived during the same age. Dinosaurs walked upright, with their legs supporting them beneath their bodies, like many mammals and birds. This feature is unique to dinosaurs and not seen in other reptiles. Lizards, for example, have sprawling limbs. Some early archosaurs had postures that were halfway between birds and reptiles.

dinosaurs were coelophysis, fast-moving, 10-foot (3 m) long flesh-eaters that hunted in packs *(below)*. They roamed across what is now the southern United States preying on insects, lizards, small mammals, and even other small dinosaurs.

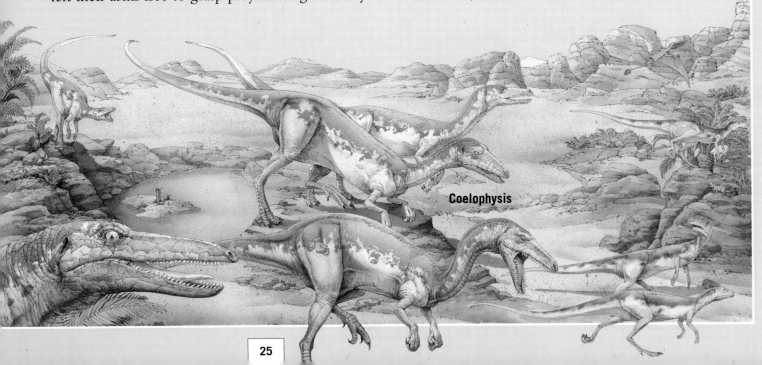

Coelophysis

DINOSAUR FAMILIES

There were hundreds of dinosaur species. They formed two major groups: the lizard-hips, or **saurischians**, had hip bones shaped like those of other reptiles. The bird-hips, or **ornithischians**, had hip bones shaped like those of modern birds. The carnivores (theropods) and the long-necked herbivores (sauropods and prosauropods) were saurischians, while the ornithopod, armored, plated, and horned *(see page 29)* dinosaurs were ornithischians.

The **sauropods** were among the largest animals that ever lived. Some species grew to more than 50 feet (15 m). They had small heads, long necks, thick, pillar-like legs, and long tails. Moving in herds, the sauropods stripped leaves from the trees with their teeth. If attacked, they might have been able to rear up on their hind legs and bring their weight crashing down on their enemy. Their tails could have been used like whips.

Brachiosaurus—the heaviest dinosaur of all—was a sauropod. The longest dinosaurs were also sauropods: the diplodocids apatosaurus, barosaurus, and diplodocus itself.

The **theropods** (flesh eaters) varied greatly in size, but most had powerful jaws and sharp teeth. Large theropods, such as tyrannosaur *(see page 29)* may have stalked their prey, bringing them down after a short chase. Smaller, swifter carnivores probably hunted in packs.

**Plateosaurus
(prosauropod)**

Brachiosaurus

Stegosaurus

The plated dinosaurs, such as stegosaurus, were large and slow-moving. They had double rows of bony plates that ran the length of their backs. They also had long spines on their tails, which they may have used to fend off attackers.

Ankylosaurs, the armored dinosaurs of the Cretaceous period, were built like tanks. Their massive bodies were encased in thick slabs of bony armor, reinforced by studs and spikes. They would become vulnerable only if they were somehow turned over, exposing their soft underbellies.

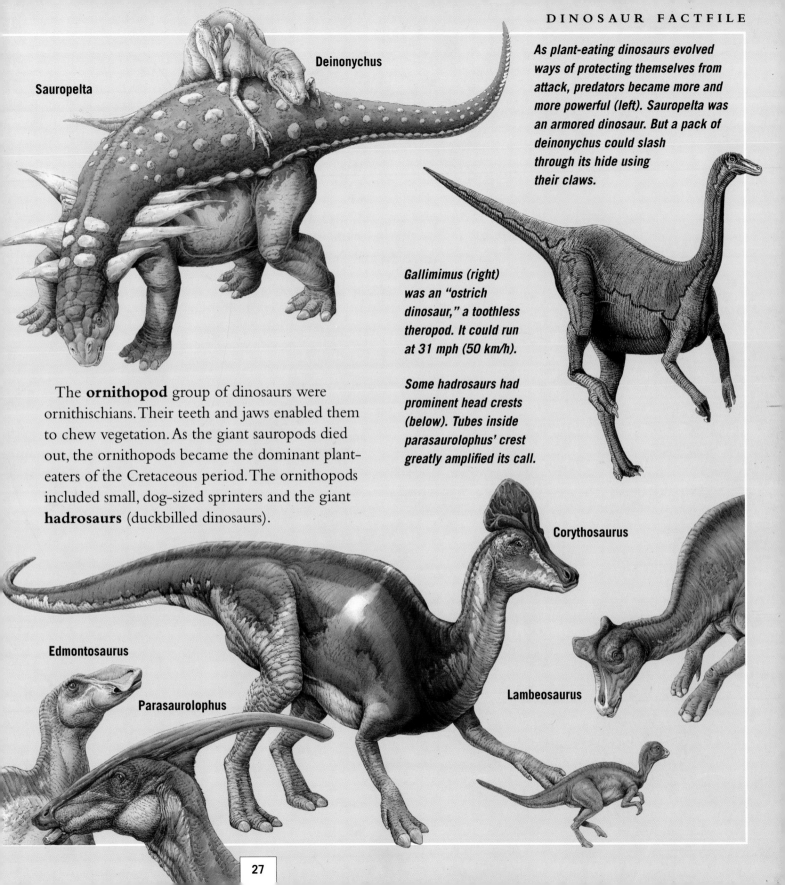

Sauropelta

Deinonychus

As plant-eating dinosaurs evolved ways of protecting themselves from attack, predators became more and more powerful (left). Sauropelta was an armored dinosaur. But a pack of deinonychus could slash through its hide using their claws.

Gallimimus (right) was an "ostrich dinosaur," a toothless theropod. It could run at 31 mph (50 km/h).

Some hadrosaurs had prominent head crests (below). Tubes inside parasaurolophus' crest greatly amplified its call.

The **ornithopod** group of dinosaurs were ornithischians. Their teeth and jaws enabled them to chew vegetation. As the giant sauropods died out, the ornithopods became the dominant plant-eaters of the Cretaceous period. The ornithopods included small, dog-sized sprinters and the giant **hadrosaurs** (duckbilled dinosaurs).

Corythosaurus

Edmontosaurus

Parasaurolophus

Lambeosaurus

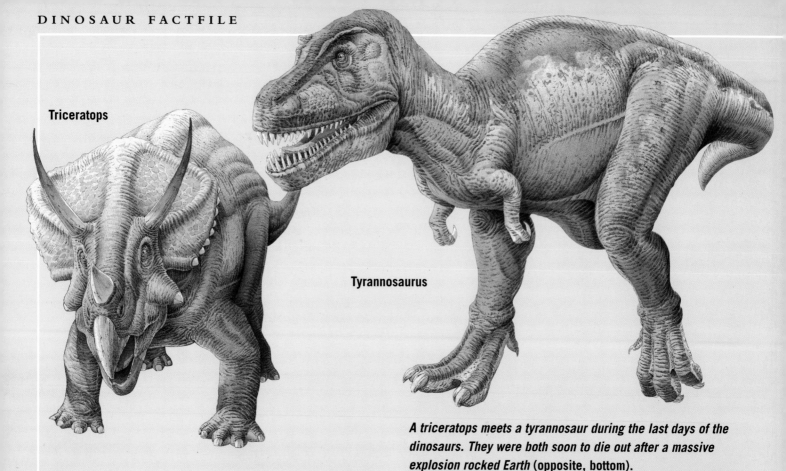

Triceratops

Tyrannosaurus

A triceratops meets a tyrannosaur during the last days of the dinosaurs. They were both soon to die out after a massive explosion rocked Earth (opposite, bottom).

DINOSAUR EXTINCTION

By the end of the Cretaceous period—about 65 million years ago—all the dinosaurs were extinct. They had ruled Earth for more than 160 million years (modern humans have existed for just 150,000 years). No one knows what happened, but evidence shows that extinction was quite abrupt. Many scientists think that a massive **asteroid**—a large rocky object in space—may have crashed into Earth *(right, above)*. This theory holds that the asteroid created a large crater and threw huge quantities of pulverized rock high into the atmosphere, blocking sunlight and lowering temperatures for many years. Another theory is that a massive **volcanic eruption** *(right, below)* could have produced the same effect.

No dinosaur, dependent as they were on warmth and a plentiful supply of food, could survive life in a cold, bleak desert.

Evidence for both theories comes from the discovery by geologists of a layer of metal, called iridium, in late Cretaceous rocks. Iridium is believed to be present only in Earth's core and in asteroids. Iridium dust thrown up by an exploding asteroid or lava from inside Earth may have settled on the surface. Millions of years later, the iridium became compacted in the rocks of the time.

The final years of the dinosaurs produced some spectacular species. They included the 30-foot (9 m) long giant triceratops, a horned dinosaur with a huge skull, a massive bony neck frill, three horns, and a parrot-like beak. This plant eater's worthy defenses would have been tested in confrontations with tyrannosaur, a massive meat-eating theropod with a huge head and powerful back legs.

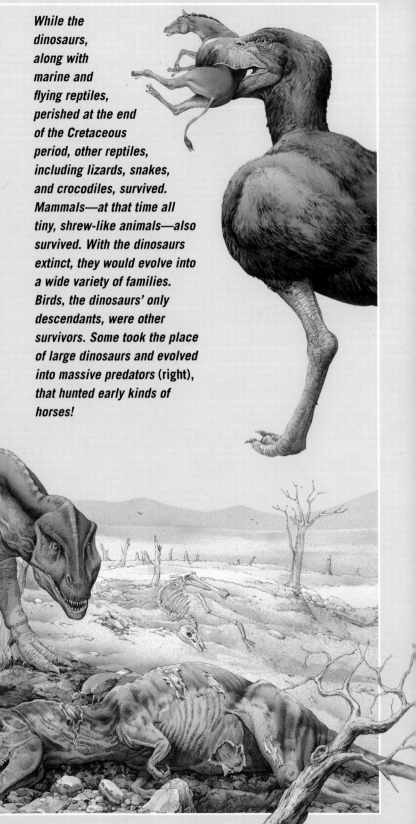

While the dinosaurs, along with marine and flying reptiles, perished at the end of the Cretaceous period, other reptiles, including lizards, snakes, and crocodiles, survived. Mammals—at that time all tiny, shrew-like animals—also survived. With the dinosaurs extinct, they would evolve into a wide variety of families. Birds, the dinosaurs' only descendants, were other survivors. Some took the place of large dinosaurs and evolved into massive predators (right), that hunted early kinds of horses!

GLOSSARY

Amphibians Animals that live much of their lives on land, but which have to return to water to breed.

Ankylosaurs Ornithischian dinosaurs fully covered in armored plates, studs, and spikes. Some, for example euoplocephalus, had tail clubs.

Archosaurs A group of reptiles that first appeared in the late Permian period and gave rise to the crocodiles, pterosaurs, dinosaurs, and birds.

Asteroid A rocky body that orbits the Sun. Asteroids range in size from tiny specks to just under 620 miles (1,000 km) in diameter.

Continents The great land masses, such as Asia, Africa, and the Americas, that make up the land surface of Earth.

Continental drift The movement of continents around the globe. Earth's outer layer is made up of separate interconnecting pieces, called **tectonic plates**, which are constantly grinding into, away from, or alongside one another, taking continents or parts of continents with them.

Dinosaurs Reptiles that lived on land during the Mesozoic era (250 to 65 million years ago) and which walked upright on legs held beneath their bodies, like birds and many mammals.

Diplodocids Sauropod dinosaurs with long, slender bodies and tails. They included apatosaurus, seismosaurus, and diplodocus.

Evolution The process by which forms of life have changed over millions of years, gradually adapting to make the best use of their environment.

Hadrosaurs "Duckbilled" dinosaurs from the late Cretaceous period. Grazing in herds, they were plant eaters with special grinding teeth.

Horsetails Plants that grew in the great swamp-forests to heights of 50 feet (15 m) or more. They have regular whorls of spiky branches.

Ornithischians The "bird-hipped" dinosaurs, one of two major types of dinosaur (the others were the saurischians). Ornithischians had backward-slanting pubic bones—the lower part of the hip bone.

Ornithopods Ornithischian dinosaurs that had teeth and jaws enabling them to chew vegetation. The ornithopods ("bird feet") included iguanodon, hypsilophodon, and the hadrosaurs.

Pelycosaurs A group of reptiles, some of which had large sails of skin supported by bone projecting from their backs.

Predators Animals that prey on others.

Prosauropods The first plant-eating dinosaurs, emerging in late Triassic times. Some may have been partly bipedal (moving on two feet).

Pterosaurs Flying reptiles that existed from the late Triassic to late Cretaceous periods. Their wings were formed from skin flaps between the fourth finger and lower body.

Saurischians The "lizard-hipped" dinosaurs, one of two major types of dinosaur (the others were the ornithischians). Saurischians had forward-jutting pubic bones—the lower part of the hip bone.

Sauropods Long-necked, four-legged, plant-eating dinosaurs. They were the largest and heaviest land animals of all time.

Theropods All the flesh-eating saurischians.